Caves

Erinn Banting

WEIGL PUBLISHERS INC.

Published by Weigl Publishers Inc.
350 5ᵗʰ Avenue, Suite 3304, PMB 6G
New York, NY 10118-0069
USA

Web site: www.weigl.com

Library of Congress Cataloging-in-Publication Data

Banting, Erinn.
    Caves / Erinn Banting.
    p. cm. — (Biomes)
    Includes index.
    ISBN 1-59036-436-8 (hard cover : alk. paper) —
    ISBN 1-59036-437-6 (soft cover : alk. paper)
    1. Cave ecology—Juvenile literature. I. Title. II. Biomes (Weigl Publishers)
QH541.5.C3B36 2006   577.5'84—dc22   2006001038

Printed in China
1 2 3 4 5 6 7 8 9 0   10 09 08 07 06

Project Coordinator
Heather Kissock

Designers Warren Clark,
Janine Vangool

Cover description: Water
reflections showcase the rock
formations inside Seven Star
Cave, located in the Guangxi
province of China.

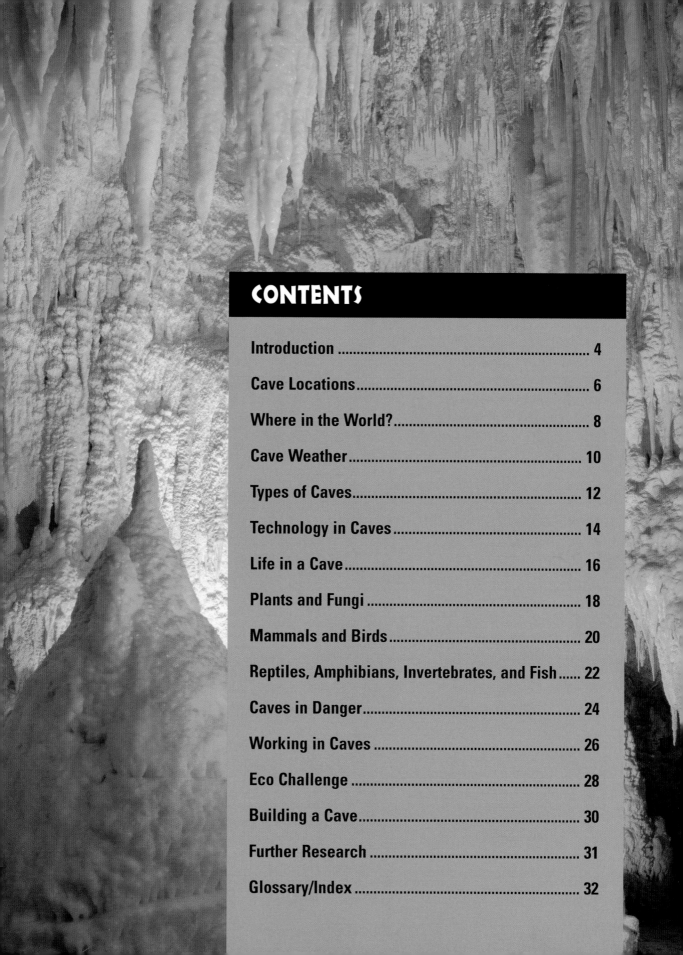

CONTENTS

Introduction

Earth is home to millions of different **organisms**, all of which have specific survival needs. These organisms rely on their environment, or the place where they live, for their survival. All plants and animals have relationships with their environment. They interact with the environment itself, as well as the other plants and animals within the environment. This interaction creates an **ecosystem**.

Different organisms have different needs. Not every animal can survive in extreme climates. Not all plants require the same amount of water. Earth is composed of many types of environments, each of which provides organisms with the living conditions they need to survive. Organisms with similar environmental needs form communities in areas that meet these needs. These areas are called biomes. A biome can have several ecosystems.

Xpukil Cave is found in Mexico's Yucatan district.

Caves can be found on each of Earth's seven continents, from the frozen northern and southern poles to the tropical regions near the equator. They can be located under mountains, in the ocean, on islands, and even inside **glaciers**.

Caves are made up of small and large **caverns**, deep pits, underground waterways, and strange-looking rock formations. They provide a natural hiding place for natural wonders and animals seeking shelter. Some animals, such as bears, use caves for warmth, shelter, and safety throughout their winter **hibernation**. Caves have also been used throughout history to hide and protect people, precious goods, and treasure hidden by smugglers or pirates. In some parts of the world, people have even built homes, storage areas, and places of worship inside caves.

Their underground location gives caves one of the most unique climates in the world. Rain never falls in caves, but rain above ground influences the flow of underground streams, rivers, lakes, and even waterfalls.

FASCINATING FACTS

Cultures around the world have different myths about the magical and spiritual significance of caves. The Japanese believed that their royal family was born from the Sun goddess, Amaterasu. This goddess brought light and prosperity to the land when she came out of her cave, where she had been hiding from her havoc-wreaking brother. In Ireland, some people believed that a cave in Connaught, on the west side of the island, was home to monsters and demons that were only able to escape on the night of Halloween to frighten people and cause mischief.

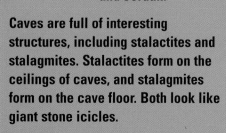

Throughout history, caves have been used to hide important goods and materials. Between 1947 and 1956, hundreds of clay jars containing text from the Old Testament were found in a network of caves near the Dead Sea, between Israel and Jordan.

Caves are full of interesting structures, including stalactites and stalagmites. Stalactites form on the ceilings of caves, and stalagmites form on the cave floor. Both look like giant stone icicles.

Cave Locations

C aves lie beneath the ground of every continent of the world. Cave systems extend for great distances and to extreme depths. Some caves even exist beneath towns and cities.

Most caves are located in karst regions. Karst refers to ground that is largely made up of limestone. Limestone is a soft stone that formed millions of years ago from the remains of sea creatures. Scientists estimate that nearly 10 percent of Earth's land surface is made up of karst.

Many of the world's caves formed over millions of years, as water and wind wore away at the karst and dug out underground tunnels and caverns. Some cave networks are very small, while others are enormous.

Limestone cliffs on the coast of Portugal have been carved out by a constant barrage of ocean waves. These cliffs contain many picturesque caves.

Mammoth Cave, which is located in Kentucky, has more than 360 miles (580 kilometers) of explored passageways. This is about three times longer than any other cave system in the world. In fact, geologists estimate that there could still be as much as 400 miles (644 km) of unexplored passages in this system.

Sarawak Chamber is the largest known cavern in the world. It is part of Good Luck Cave, which is part of the Mulu cave system in Malaysia. The cavern is 2,297 feet (700 meters) long, 328 feet (100 m) tall and 1,312 feet (400 m) wide. This is big enough to hold more than 7,500 buses.

Some cave systems do not have deep caverns, but are made up of long passageways. In central Asia, the main tunnel of the Boj-Bulok Cave runs for more than 4,642 feet (1,415 m), but is only an average of 20 inches (0.5 m) wide.

FASCINATING FACTS

Scientists believe there is a cave system in Italy that may be as deep as 5,900 feet (1,800 m). That is deeper beneath the Earth than any human being has ever traveled.

Mammoth Cave was discovered in 1798. Mummies and rock paintings thousands of years old have been discovered within the network's walls.

WHERE IN THE WORLD?

This map shows where some of the world's cave systems can be found. Find the place where you live on the map. Is it above a cave or cave system? If not, where is the closest cave or system to you?

Arctic Ocean

Asia

Europe

Boj-Bulok Cave

Mogao Caves

Pacific Ocean

Africa

Atlantic
Ocean

Mulu Caves

Ankarana Caves

Australia

Indian
Ocean

Waitomo
Glowworm Caves

Cave Weather

C aves are often believed to be cold and damp places, but this is not always the case. Whether they are hot or cold, dry or wet depends on their location and the environment around them. Sometimes, the inside of a cave can have warmer temperatures than the area outside of the cave.

Heat can enter a cave in four ways. Wind can blow heat in through the cave opening. Water flowing into the cave can also bring outside temperatures with it. Rocks lying above the cave can provide **insulation** and help keep the inside warm. Rocks below the cave can also bring warmth as a result of the hot **magma** that bubbles beneath Earth's crust.

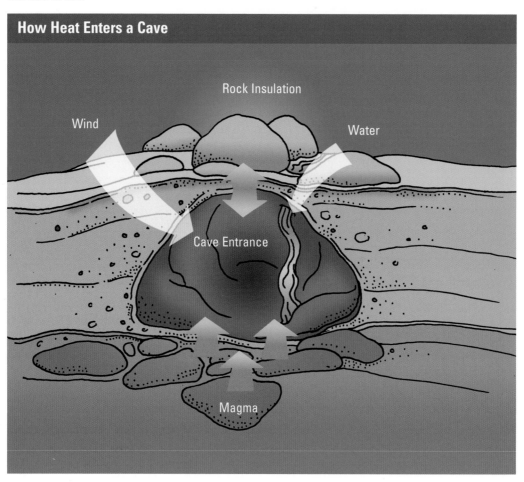

How Heat Enters a Cave

Rock Insulation

Wind

Water

Cave Entrance

Magma

Water runoff from storms and melting snow often drains into cave systems. Flash flooding can occur as a result, filling the cave with water in very little time.

Outside weather can affect more than just the cave's temperature. **Humidity** levels can be affected as well. This occurs when cold air comes into the cave. Cold air is normally dry, and cave air tends to be moist. When the cold air enters the cave, it dries the air inside, making it less humid. As the moisture evaporates, the air becomes cooler, and the temperature inside the cave lowers.

External **precipitation** and wind can affect the inside of a cave on a grander scale as well, depending on where in the world the cave is located. In areas that receive large amounts of rain, such as parts of South America and Asia, caves sometimes flood during the monsoon, or rainy, season. Wind also whistles and blows through some caves, affecting the air pressure inside. When the air pressure outside of a cave increases, more air is forced into the cave, increasing the air pressure inside. When the air pressure outside drops, some of the air inside the cave flows back out, reducing the air pressure in the cave as well. A slight breeze develops in the cave as the air rushes in and out. This process helps the air circulate within the cave.

Types of Caves

Caves develop in a variety of ways. They have been one of Earth's landforms for thousands, even millions, of years. However, some caves are more recent in their development. Still others are more temporary in nature, relying on the climate for their longevity.

Sandstone Caves

Millions of years ago, shallow seas covered parts of the planet. Over time, the water dropped, and the sand that had once covered the sea floor was pressed into solid rock called sandstone. Water from rivers and lakes slowly **eroded** and carved out the chambers and tunnels that became sandstone caves.

Water and wind erosion created the graceful curves found in the sandstone of Arizona's Grotto Cave.

Caves are still forming in the limestone of Nevada's Lehman Caves.

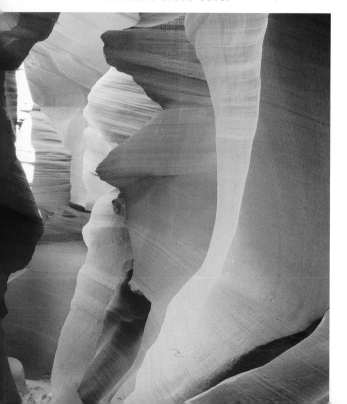

Limestone Caves

Long ago, millions of organisms lived in the shallow seas that covered Earth. When these organisms died, their bodies sank to the bottom of the seas and decomposed, or broke down. The weight of the water caused these shells and bones to form hard layers of rock called limestone. As Earth evolved, the limestone rose above the ground, where it was carved out further by wind and water. Limestone caves resulted from this process.

The rocky coast of California's Pfeiffer Beach features several sea caves.

Sea Caves

Deep beneath the sea, powerful currents and waves wear rock away to form underwater caves. Waves slowly erode rock from cliff surfaces until pockets form at their bases. As water seeps into these pockets, caves continue to grow. Some sea caves have areas that are above water and other parts that are submerged.

Lava-Tube Caves

When a volcano erupts, **molten** lava runs down its sides. Because the lava is so hot, its surface cools before its interior does. The surface hardens as it cools, forming large tunnels or tubes beneath. The molten lava continues to run through these tubes until the eruption stops. The tubes then remain until the next eruption.

More than 500 lava-tube caves can be explored at California's Lava Beds National Monument.

Alaska's Muir Glacier contains the Blue Ice Cave, named for its icy-blue appearance.

Ice Caves

Ice caves are created by the movement of glaciers. As glaciers move across the ground, they melt. Meltwater runs beneath them and hollows out caves inside the glaciers. These caves form and melt very quickly.

FASCINATING FACTS

A cave can be divided into three zones. The entrance zone contains the passage to the surface. This zone receives the most light of any zone. The twilight zone is located a bit deeper in the cave. Light still reaches this zone, but not as much as the entrance zone. The dark zone is the deepest part of the cave. It is so far underground that no light reaches it.

Technology in Caves

Studying caves helps scientists understand how Earth has grown and changed over time. By analyzing the development of caves and the life forms found inside, scientists can gain a better understanding of this unique biome.

The study of caves is called speleology. Speleology is a branch of geology. This is the study of Earth's formations and composition, and the processes that shape it. Speleology combines geology with the study of other branches of science. One branch is mineralogy, or the study of gems and minerals. Another is hydrology, the study of how water shapes the land. Speleology also combines aspects of biology, the study of living organisms.

Speleologists must be good at caving, or exploring a cave. Cavers are skilled at moving through a cave's narrow passages, giant caverns, and sometimes dangerous waterways. Cavers must also be able to climb up and down steep surfaces. They use the same equipment as rock climbers, including strong nylon rope, an **ascender**, and a rappel rack, which controls how fast a person moves down a cliff's side. When exploring dark zones, cavers wear a hard hat that shines a bright light into the area.

A scientist uses standard caving equipment to rappel through one of the numerous openings of the roof of Cueva de Villa Luz (Cave of the Lighted House) in Mexico.

While collecting a water sample in caves, geologists sometimes wear a gas mask to protect themselves from the toxic gas fumes that may dwell within the cave.

Once inside the cave, speleologists begin their studies using specialized tools. These may include measuring tapes and clinometers, which scientists use to measure angles in caves they are surveying. **Chocks** and **pitons** also help speleologists maneuver through the underground world. One way speleologists study caves and cave networks is through dye gauging. Dye gauging involves releasing dye into a river in a cave to monitor where the water flows. Dye gauging helps determine the size and complexity of a cave system. Advanced methods of dye gauging use leucophor. This clear liquid can be detected under phosphorescent lights so that speleologists can track a river's direction even in the darkest parts of a cave.

FASCINATING FACTS

Early explorers who ventured into caves used crude methods to track where they had gone. To make sure they did not get lost, some explorers ran string behind them so they could find their way back out.

In 1821, Austrian researcher Adolf Schmidl made the first accurate maps of caves. Some call him the first speleologist. Throughout his career, Schmidl mapped many caves around the world.

Speleologists sometimes spend months at a time beneath the ground studying caves.

LIFE IN A CAVE

Caves may not seem like the most appealing places to live, but they are home to hundreds of crawling, leaping, and swimming creatures. Caves provide shelter and protection from the weather and predators. They also provide a safe place for some animals to hibernate during the long winter months.

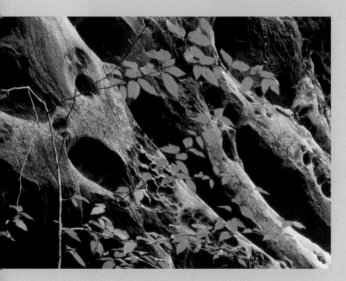

A variety of plants grow along the wall of Hazard Cave in Tennessee.

PLANT LIFE

Despite the lack of sunlight and water, caves have plant life of their own. The type of plants found depends entirely on that cave's surroundings, where it is located, and the climate of the region it is in. Most plant life can be found at the entrance to a cave, where sunlight reaches. Caves often have mosses, algae, and low-growing plants that require little sunlight to grow. No plants or trees grow in the dark zone of a cave because plants depend on sunlight to survive.

AMPHIBIANS AND REPTILES

Amphibians are cold-blooded animals that are born in water. Reptiles are also cold-blooded, but are born on land. Both use the temperature of their surroundings to regulate the temperature of their bodies. Reptiles can be found in the entrance and twilight zones of caves, and include various species of snakes. Amphibians can be found in all three cave zones. The most common amphibian found in the cave biome is the salamander.

During the day, cave salamanders live in moist, damp places near the entrance to caves. They come out at night to feed.

There are 39 cave crayfish species in North America.

INVERTEBRATES

Invertebrates are animals that do not have a spinal column, or backbone. These include insects, **arachnids**, and **crustaceans**. Invertebrates live in all of the cave biomes zones. Some examples are moths, crickets, spiders, and crayfish.

FISH

Though many caves only have shallow pools of water in them, these pools often contain fish and other aquatic life. Caves carved out by the oceans or seas often have tidal pools in their entrance or twilight zones. Waves wash and pull sea creatures into and out of the cave. Some caves also have marine life in their dark zones. One type of fish commonly found in the cave biome is the blind cavefish.

Fish can often be found swimming in and around sea caves.

MAMMALS

Mammals inhabit the entrance and twilight zones of the world's caves. Small mammals, such as mice, rats, voles, and raccoons, move between the shelter of a cave and the world beyond. Large mammals, such as bears, frequently use caves to hibernate during winter. The most common mammal found in the twilight zone is the bat. Hundreds of bat species live in caves around the world, including gray, brown, and pipistrelle bats.

BIRDS

There may not seem like many places in a cave for birds to swoop and soar, but many species find shelter and build nests in the entrance and twilight zones of caves. Birds such as owls, swiftlets, and phoebes make their nests on ledges along the cave walls.

Plants and Fungi

Cave Plants

Plants depend on light to grow because of a process called photosynthesis. Through this process, plants convert sunlight to energy. As caves have very little light, very few plants grow in caves. Mosses and ferns are the most common plants found in caves. They grow in the damp entrance zone. Mosses grow well in cave biomes because they do not have roots. Instead, they attach themselves to rocks with rootlike structures called rhizoids.

Fungi

Fungi are other organisms found in the cave biome. They thrive in dark, damp caves because they do not need light to grow. Fungi also grow well in caves because the items that they feed on are often present. Dead animals, droppings, and the dead leaves, stems, and twigs carried by underground streams can all be found in caves. Mold is the most common type of fungus that grows in caves. Mushrooms, another type of fungus, are sometimes found in caves as well.

Mosses can often be found growing in the cracks of cave walls.

Lichens are often found growing on rock.

Lichens

Lichens grow in most caves. Like mosses, lichens have no roots, so they can grow on rocky surfaces in caves. Lichen comes in many colors, including red, yellow, and orange, but most commonly it is brown or gray. One type of lichen that grows in caves is crustose lichen. This lichen grows close to the ground, is hard to the touch, and has a scaly appearance. Foliose is another type of lichen that thrives in caves. It grows in leaflike shapes.

Mammals and Birds

Sheltered Mammals

Some mammals make their way into and out of the entrance zone of caves, finding shelter and hibernating during winter. In North America, the largest mammal to hibernate in caves is the black bear. Other small mammals, such as mice, rats, raccoons, and skunks, also inhabit the entrance zones of caves. In more tropical climates, big cats, such as jaguars, take shelter in cave biomes.

Raccoons sometimes make their home in caves that are near water.

Nectar-feeding bats roost in the Aripo Caves of Trinidad during the day.

Breathtaking Bats

The most common type of mammal found in the cave biome is the bat. Hundreds of species of bats live in caves around the world. Some take shelter and sleep in the caves during the day and leave to hunt at night. Others never leave. Bats can find their way around in dark caves because they use sound to help them detect where they are going. They project high-pitched sounds that bounce back from an object to let them know it is there. Fruit bats are one of the most common types of bats to find shelter in caves, especially in tropical climates. They roost, or hang, in caves during the day and leave at night in search of fruit to eat.

Bashful Birds

Two species of birds commonly found in caves are the swiftlet and barn owl. These birds build their nests on the cliff surfaces of the twilight zone of caves throughout North America and Europe. The Guacharo, or oilbird, is a type of bird commonly found in the caves of South America and the Caribbean. These rare birds have wingspans of up to 3 feet (0.91 meters). Guacharos are nocturnal, which means they are active at night.

Barn owls leave their caves at night to hunt for food.

FASCINATING FACTS

Like bats, oilbirds use sound to help them navigate and find food.

Vultures also live in some caves. Vultures prey on the meat of animals that have been killed by other animals, humans, or vehicles.

There are nearly 1,000 species of bats in the world. Most bats eat insects, fruit, small animals, or fish, but some eat other bats.

Reptiles, Amphibians, Invertebrates, and Fish

In nature, box turtles may live to 80 years of age.

Reptiles

Snakes and turtles are common inhabitants of the entrance zone of many caves. Snakes such as copperheads and rattlesnakes live and hunt in caves. Box turtles, which get their name because of their hinged shell, also find refuge in the cave biome. Other reptiles that are not normally found in caves, such as crocodiles, have been discovered in the Ankarana Caves in Africa. Scientists believe that one of the reasons that crocodiles **adapted** to cave-dwelling was to escape being hunted by humans.

Amphibians

Amphibians enjoy the cool temperatures and damp conditions that caves provide. Salamanders are common cave dwellers. The redback salamander is gray in color, with a red or orange stripe running down its back and tail. It is often found near cave entrances. To escape from danger, this salamander coils up and tucks its tail under its head. It then wiggles its tail to distract the predator. If part of the tail is removed by the predator, it will grow back. Texas blind salamanders have colorful red gills around their heads, but look like they have no eyes. In fact, their eyes are beneath their skin, so they cannot see. Scientists believe the salamanders once had eyes, but adapted to their dark environment, so they did not need sight.

Besides caves, salamanders can also be found under logs or rocks, as well as near streams and other moist areas.

Blind Cavefish

Like blind salamanders, some species of fish adapted to living in the dark zone of the cave biome. Blind cavefish are the most common type of fish found in the caves of North and Central America. Like blind salamanders, some blind cavefish have eyes beneath their skin. Still, many blind cavefish have no eyes at all. They use water vibrations to detect objects and prey in front of them as they swim through the water.

Glowworms

In the Waitomo Glowworm Caves of New Zealand, the dark zones are often illuminated with what looks like a sky full of stars. Glowworms have glands that emit light on their abdomen. These invertebrates use this light to attract prey. Once they have attracted prey, they use their long feeding lines to catch it. Glowworms can have up to 70 feeding lines, which are like a spider's web, to trap insects.

FASCINATING FACTS

Cave crickets are common around the world. Cave crickets in New Zealand and Australia are called cave wetas. When they run low on food, these invertebrates sometimes eat their own limbs.

Blind cavefish have adapted to living in the dark in other ways. They have no pigment, or color, in their skin because it is never exposed to sunlight.

Some glowworm feeding lines are 8 inches (20 cm) long.

Caves in Danger

Scientists work hard to find ways to protect Earth's biomes. The greatest threat caves face is development by humans. Caves have a delicate ecosystem that contains unique animals, some of which can be found in no other biome in the world. Caves also contain rare and delicate rock formations, ancient fossils, and rock paintings that give us clues to the past, as well as valuable information about how Earth grows and changes.

Mining poses a large threat to the cave biome. Drilling for natural gas and oil, and digging for minerals, such as gold, copper, and silver, removes the rock that forms caves and may destroy the caves themselves. Industries are constantly looking for new sources of oil and minerals. Their actions have forced many governments to pass laws protecting cave habitats.

Mining can leave sediments in caves that clog cracks, alter water drainage patterns, and change the cave's humidity. It can also disrupt airflow in the cave.

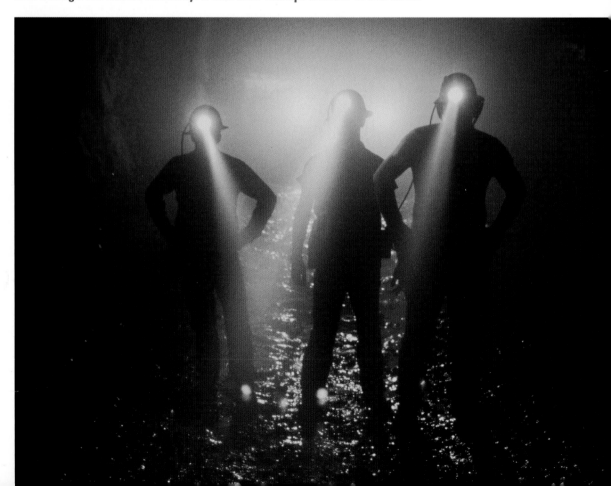

The unique plant and animal life that depends on water for survival is also threatened by pollution. Pollution from the increasing number of cities on Earth seeps into the water that eventually flows underground through caves. Amphibians breathe oxygen through their skin. If there are pollutants in the water, the animals also absorb these in their skin. These pollutants may impair the animal's ability to swim, catch food, and reproduce. Reptiles can experience similar problems. When exposed to harmful pollutants, they may produce eggs with thinner shells and have fewer young. Plants also rely on water for survival and are equally affected by pollution. Many species of fern and lichen are now protected in caves around the world as endangered species.

Tourism has also had a negative impact on caves. In order to encourage visitors, bridges, elevators, and roadways have been built within many mighty caverns to make them more accessible. These constructions have sometimes weakened cave walls or disrupted the movement of water, air, or animals within them.

FASCINATING FACTS

The Mogao Caves in China contain more than 490 passageways and caverns. The caves have been declared a World Heritage Site by the United Nations Educational, Scientific and Cultural Organization (UNESCO) because they are home to some of the oldest rock paintings on Earth. Thousands of artifacts, including coins, statues, and more than 50,000 manuscripts, in many different languages, have also been found in the caves. The caves are protected because they represent an important link to the history of China.

WORKING IN CAVES

The people who study the land, plants, and animals that live in caves have an excellent knowledge of science, history, and math. A good education is needed to study the geology of a cave, the culture and customs of the people who depended on caves, and the wildlife that lives in caves. People who work in caves often study sciences such as geology, biology, anthropology, and speleology.

SCIENTIFIC SPELEOLOGIST

- Duties: collects and records data from cave environments

- Education: bachelor of science degree

- Interests: Earth sciences, cave ecosystems and wildlife, biology, geology

Speleologists study the plants, animals, and geology of caves. Much of their time is spent working underground, sometimes in laboratories set up in the caves. Scientific speleologists can specialize in many areas, including studying the effects of climate change on the cave biome.

CAVE SURVEYOR

- Duties: measures and maps cave networks

- Education: bachelor of science degree in surveying

- Interests: mathematics, physics, computer science, drafting

Cave surveyors explore cave biomes to record data such as the height, length, and width of the cave. They also map the location of main pathways, as well as any parts of the path that branch off. They spend much of their time outdoors and must be in good physical shape in order to carry their heavy surveying equipment as they explore the cave.

CAVE CONSERVATIONIST

- Duties: ensures that caves are managed and protected in a sustainable manner

- Education: bachelor of science degree, rock climbing training

- Interests: outdoors, rock climbing, Earth sciences

Cave conservationists lobby for the protection of cave biomes from human development. They do this by exploring and mapping the inside of caves, and by noting any plant and animal life living within the cave that could become endangered if the cave biome was destroyed. Cave conservationists also write proposals to suggest ways to save the cave environment.

ECO CHALLENGE

1 What are the three zones in a cave?

2 Which is the largest known cavern in the world?

3 What is speleology?

4 What is the difference between a stalactite and a stalagmite?

5 Where can ice caves be found?

6 How many feeding lines might a glowworm make to catch its prey?

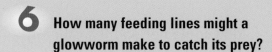

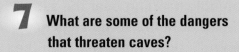

7 What are some of the dangers that threaten caves?

8 What is another name for the oilbird?

9 What do mosses use to attach themselves to rocks?

10 What are some of the interests a cave surveyor might have?

Answers

1. the entrance zone, the twilight zone, and the dark zone
2. the Sarawak Chamber in Malaysia
3. a branch of geology that focuses on the study of caves
4. Stalactites form on the ceilings of caves, stalagmites form on cave floors.
5. inside glaciers or icebergs
6. up to 70
7. mining, pollution, and tourism
8. the guacharo
9. rhizoids
10. physics, mathematics, drafting, computer science

BUILDING A CAVE

Cave formations such as stalactites and stalagmites grow over thousands, sometimes millions, of years. In this activity, you can watch them build and grow in your own home or classroom.

MATERIALS

- 3 saucers
- 6 jars or cups of the same size
- Epsom salts
- warm water
- 3 pieces of yarn or string
- baking soda (sodium bicarbonate)
- washing soda (sodium carbonate)

1. Place one saucer between two of the jars or cups.

2. Dissolve as much Epsom salt as you can in warm water in each of the containers.

3. Soak a piece of string in the solution. Then, put one end in one container and the other in the second container with the middle hanging over the saucer. Make sure the string sags in the middle until it is lower than the water level in the two containers.

4. Repeat this process with the other saucers and jars using baking soda and washing soda and the other two pieces of string.

5. Check your experiment each day. What sorts of formations grow on each of the strings? Are there differences between them? Which ones look like formations you might see in a cave?

FURTHER RESEARCH

How can I find more information about ecosystems, caves, and animals?

- Libraries have many interesting books about ecosystems, caves, and animals.

- Science centers, museums, and aquariums are great places to learn about cave life under the ground and sea.

- The Internet offers some great websites dedicated to ecosystems, the world's biomes, and caves.

BOOKS

Aulenbach, Nancy Holler. *Exploring Caves: Journeys into the Earth*. Washington, DC: National Geographic Children's Books, 2001.

Lindop, Laurie. *Cave Sleuths*. Washington, DC: 21st Century Books, 2004.

Taylor, Michael Ray, and Ronal C. Kerbo. *Caves: Exploring Hidden Realms*. Washington, DC: National Geographic, 2001.

WEBSITES

Where can I learn more about caves and other biomes?

Encarta Encyclopedia
www.encarta.com

Where can I go on a virtual cave tour?

The Virtual Cave
www.goodearthgraphics.com/virtcave/

How can I learn more about ecology?

Kids Do Ecology
www.nceas.ucsb.edu/nceas-web/kids/ecology/faq.html

GLOSSARY

adapted: changed to fit an environment

arachnids: a group of animals, including spiders, that are related to insects

ascender: equipment that prevents slipping down a rope

caverns: large caves that are mostly underground

chocks: climbing equipment that anchors the climber to the rock

crustaceans: animals with segmented bodies, jointed limbs, and an outer shell

ecosystem: a community of living things sharing an environment

eroded: worn away

glaciers: large, slow-moving bodies of ice

hibernation: the act of passing the winter in a resting state

humidity: moisture in the air

insulation: material that is used to slow or stop the flow of electricity, heat, or sound

magma: melted rock below Earth's surface

molten: melted

organisms: living things

pitons: metal spikes hammered into a rock face to anchor a climber

precipitation: rain, snow, or hail

INDEX